THE SECRET
LIVES of PUFFINS

THE SECRET LIVES OF PUFFINS

Photographs by Mark Sisson
Text by Dominic Couzens

BLOOMSBURY
LONDON · NEW DELHI · NEW YORK · SYDNEY

Published 2013 by Bloomsbury Publishing Plc,
50 Bedford Square, London WC1B 3DP

ISBN (print) 978-1408-1-8667-1
A CIP catalogue record for this book is available
from the British Library.

This book is produced using paper that is made from wood grown
in managed sustainable forests. It is natural, renewable and
recyclable. The logging and manufacturing processes conform to
the environmental regulations of the country of origin.

Page design and typesetting by Nimbus Design

Printed in China by C&C
10 9 8 7 6 5 4 3 2 1

Acknowledgements

Mark Sisson Although wildlife photography as a profession
is by its very nature often a solitary business, there are
several people and organisations I need to thank for their
help in making the portfolio of images in this book a reality.

The first of these is my wife Caroline who, along with the
rest of my family, puts up with my long absences in the pursuit
of my work and supports me in the process. Thanks to all who
have joined me on a number of the trips involved, specifically
fellow photographers Danny Green and Paul Hobson who have
also provided advice and inspiration at the times when it was
needed, and Kevin Treadwell who has helped me out when
additional photographic kit became a necessity beyond my
budgets! Also all involved with the management of Skomer
Island and Fair Isle Bird Observatory where over a number of
years I have stayed and spent many days and weeks with your
puffin colonies. Finally to those who manage the conservation
at the many other colonies in the UK, Iceland, Ireland and
Norway where I have done the same: without your work the
many pressures on puffins would have a greater toll.

Dominic Couzens I thank my wife Carolyn and children
Emmie and Sam for keeping me sane while earning a living
as a writer. Thanks to Lisa Thomas for believing in this
project, and to Mark Sisson for re-establishing a friendship
that began at school.

CONTENTS

INTRODUCTION

This book is a showcase of the intimate life of the Atlantic Puffin, as seen through the lens of wildlife photographer Mark Sisson. Over many years, Mark has worked on puffins in many parts of their range, and this collection, covering many facets of the breeding season life of this bird, includes images from Iceland, Norway, England, Scotland, the Republic of Ireland and Wales. Doubtless many of those readers leafing through the book will recognise one or more locations, perhaps including those where they first set eyes on the bird in the wild.

The accompanying text is intended to cover the details of the puffin's biology, either through captions or through the introductory paragraphs to each chapter. Over the pages it is hoped that readers will get quite a full picture of how this bird lives: how it feeds, how it acquires a mate, how it constructs its nest, feeds its young and so on. Not everything that the puffin does can be captured on film, so the book mainly deals with what a visitor to a puffin colony might reasonably see for themselves. Details such as the subterranean life of young puffins, or the behaviour of the adults when they are fishing out to sea, are not illustrated, but are described in the longer blocks of text that head each chapter.

The primary purpose of the book is to educate puffin fans by presenting images of this bird's life and explaining what is going on. Hopefully, though, the images will also inspire people to get out and see puffins for themselves, for there is no substitute for experiencing them in real life. This remarkable little seabird is also one among many strong marine characters in our avifauna, and all of them need their habitats, including the open sea, to be looked after and preserved. Being so popular, the puffin draws the public into a love of seabirds and wild places. If this book helps to enhance that process, it will have done what it set out to do. ▪

THE PUFFIN

The Atlantic Puffin (*Fratercula arctica*) is a small seabird with an outsize, brightly coloured bill and orange legs. It belongs to a family known as the Auks (*Alcidae*), along with 20 or more other species, of which five others, the Guillemot (*Uria aalge*), Razorbill (*Alca torda*), Black Guillemot (*Cepphus grylle*), Little Auk (*Alle alle*) and Brünnich's Guillemot (*Uria lomvia*) also breed in Europe, often in the same locations as puffins. Its world breeding range encompasses the North Atlantic coasts of Europe and North America, hence the name. There are two other species of puffin, the Horned Puffin (*Fratercula corniculata*) and the Tufted Puffin (*Fratercula cirrhata*), both of which occur on the Pacific coast of North America and Asia. They are not covered in this book.

For a seabird, the Atlantic Puffin is diminutive. It stands just 18–20cm tall (7–8in), which is about as tall as a paperback novel, and it is 26–29cm (10–11.5in) long from the tip of its bill to the end of its short, pointed tail. By comparison, a guillemot is 46cm (18in) long and a herring gull up to 60cm (24in) long. The puffin weighs between 350 and 600g, while the herring gull weighs up to 1500g. Seeing these figures it is easily understandable how puffins can be bullied by herring gulls and other birds.

In terms of colour and pattern, almost everybody knows what a puffin looks like. If you stop to think about it, this fact is quite surprising, because not many people have ever actually seen a puffin in the flesh. It lives during the summer in remote places that are often hard to get to, and in the winter it disappears out to sea. So it isn't as though the puffin is a well-known, back garden kind of familiar neighbour. The truth is that most of us only know it second hand. But the puffin has such a singularly unusual shape and colour that it captures the imagination.

Portrait of a puffin in breeding plumage. In common with many seabirds it is black above and white below, with the only colour on the legs and head. **[RIGHT]**

The puffin is a portly but upright bird, with a plump body, very large, rounded head and extremely short, bluntly pointed tail. The legs are quite long, so that puffins can walk and run quite well on land; their relatives the guillemot and razorbill have short legs and can only waddle at best. The tarsus (main leg bone) of the puffin is thick and rounded, while the feet have three well clawed toes joined by webs. At first sight, the most obvious feature is the remarkable bill, which is triangular in shape and, in the breeding season, spans the height of the head from crown to chin. Although large, the bill is laterally compressed, so that it looks narrow when the bird is face on. The bill is also highly coloured and unmistakable, and often described as 'parrot-like': the outer part is deep orange and the inner half dark bluish-grey, and there is a yellow chevron-shaped line dividing the two parts. There is also a yellowish fleshy plate dividing the upper mandible from the forehead and a bright, wrinkled yellow rosette of bare skin at the base of the mouth. Depending on the age of the bird (see below), there are a number of yellowish grooves on the outer part of the bill.

As far as plumage is concerned, the puffin is boldly black above and brilliant white below. The black extends from the forehead over the crown, and covers the back, the whole of the wings and the tail. There is also a black band across the chest. In late summer the black feathers wear a bit and may look browner. Below the chest the underparts are entirely clean white apart from a dusky patch on the flanks, and the underwings are dusky grey, a good distinction from razorbills and guillemots, which have glinting white under the wing. The face is greyish white, darkest on the throat. There are no plumage differences between male and female puffins, but the male is consistently slightly bigger and heavier than the female, although this distinction is not visible in the field. ⟩

As you might expect, the puffin has a broad gape. This gives it a good reach when catching fish in its jaws underwater. **[ABOVE]**

The tongue is relatively large, and orange in keeping with the rest of the gape. The tongue helps to keep the fish secure when the puffin is carrying them back to the burrow. For this purpose it is fitted with small, backward-projecting spikes. **[BELOW]**

The neck ring of this individual is narrower than that of the bird above, and there is a wavy lobe of black on the side of the breast. [LEFT]

The face isn't quite white, but actually a light ash-grey, and the precise shape of the face pattern varies quite a bit between individuals. In this bird, the grey comes to a neat point towards the back of the head. [BELOW LEFT]

This book, by the way, only showcases puffins in their breeding plumage, as described above. However, in winter puffins undergo some changes. The most obvious difference is that the bill is smaller and doesn't fit so snugly on to the head; indeed it looks a little strange and grotesque, like an outsize nose. It is still colourful, with essentially the same orange and bluish-grey combination, but not quite so bright. The difference in shape is brought about by the loss of horny sheaths on the outer part of the bill. In the spring an adult puffin develops several sheaths and plates on both mandibles, including a rim on the upper mandible and some broad orange plates covering the middle. When not needed in autumn these simply fall off, to grow again next year.

Another major difference from summer to winter is that a puffin's face becomes much darker, especially around the eye where it looks charcoal-stained, and the ornaments of spring are lost. The rest of the plumage looks the same as it does in summer.

Young puffins take four years to reach maturity and their progress can be seen in their bill grooves. When it is a year old a youngster has only the hint of a bill groove, whereas a year later it has one fully developed groove. When three years old it has one and a half grooves and on its fourth birthday it has two. It will not breed until it has two complete grooves; older birds sometimes develop as many as four grooves in all.

Standing up on two legs, with narrow forelimbs and being somewhat portly, with a large head, there is something of the human caricature about the puffin, and this undoubtedly accounts for the bird's extraordinary popularity. The majority of other bird species are too alien for us to feel any kind of empathy. Penguins share a similar body plan and they, too, are hugely popular with people. The admiration, we can assume, is unrequited.

Puffins and other auks are often described as being like 'flying penguins', and the analogy is actually perfectly valid. Although the two families are not at all closely related (auks are closer to gulls and skuas than penguins), it is certainly fair to say that each family is the ecological equivalent of the other in its respective hemisphere, auks in the Northern Hemisphere and penguins (Spheniscidae) in the south. Each family has a similar number of species, 22 auks to 18 penguins. The members of both families are mainly black above and white below, with brighter markings around the head; both nest in large and often crowded colonies; both families spend most of their time swimming; both have a basic diet of fish, but also plankton and many other sea creatures; both pursue prey underwater; both use their wings and their feet in locomotion underwater; both are entirely at home in cold waters, often around icebergs. It is fascinating to realise that none of these striking similarities have anything to do with genetic links, but are instead the result of evolution bringing about analogous body forms and lifestyles in a broadly similar environment. ›

The black band across the neck varies considerably in its thickness and pattern. This bird has a broad band with a poorly defined lower rim. Characteristics such as these help the puffins to recognise each other individually. **[RIGHT]**

In most puffins, the top of the crown is jet black, abutting the top of the bill reasonably neatly. **[BELOW]**

In some individuals, there is quite a lot of whitish feathering on the forehead. This bird also has a few whitish specks on its crown, which is not at all unusual. [LEFT]

Although the puffin's bill looks broad in side view, it is actually laterally flattened. This enables it to slice through the water with minimal resistance. [ABOVE]

As an aside, the above statement about auks being 'flying penguins' is not quite true, because until comparatively recently there was a member of the auk family that was flightless. The Great Auk (*Pinguinus impennis*) was, indeed, the most penguin-like of the auk family and its generic name *Pinguinus* is the origin of the word 'penguin'. It once shared colonies with puffins and other auks, including on St Kilda, the Orkneys, the Faeroe Islands and Iceland. It is now extinct, have been last seen in Newfoundland in 1851.

There are a few other characteristics of auks generally that are worth a mention. All the species obtain the liquid they need from their prey, and as far as is know do not drink any water, either fresh or seawater. They all have thick, dense and rather heavy plumage that keeps them warm at high latitudes. They all have 11 primaries (main flight feathers), although one is tiny and their 15–19 secondary feathers are rather short, contributing to the narrow wing shape. The tails are invariably short, usually with 12 main feathers, although puffins happen to have 16, though why they need the extra four is anybody's guess.

All birds have to change their feathers from time to time, because just as humans need a wardrobe makeover at regular intervals, wear and tear ensures that feathers degrade and need replacing. Although moult is not the sort of subject to excite birdwatchers, the process is incredibly important to the birds themselves. Furthermore, in the case of the puffin moult is a truly intriguing part of their lives, and one about which we understand very little. After breeding it seems that puffins moult their feathers in two stages, changing the body feathers in the late summer and then the flight feathers, of the wings and tail, in the late winter. However, some puffins in a few colonies have been found to be changing their flight feathers in the autumn, which raises the intriguing question of whether the moulting schedule differs between colonies, or even according to conditions. That would be a highly unusual state of affairs.

Whatever season they actually do it, the flight feather moult renders puffins flightless for about four to five weeks. This is typical for a wide range of seabirds, including the other auks, shearwaters and sea ducks, so clearly the oceanic environment is a safe place to live, even when you cannot fly. This suggests that the birds are not frequently picked off from below, although there are a few records of puffins in the stomachs of large fish.

The puffin may not be a familiar wild bird, but that does not make it rare. It has a population of about 20 million individuals, of which six to seven million pairs breed in a given year. The total breeding range encompasses 1,620,000 square kilometres, on both sides of the Atlantic Ocean. There are colonies in Russia, Norway, Sweden, Iceland, the Faeroe Islands, the British Isles, France, Greenland, the USA and Canada. Although the species is suffering a long-term decline overall, it is stable or increasing in some places, and is not considered to be globally threatened as yet. The vast majority of colonies are so isolated that habitat destruction, such a menace to so many birds throughout the world, is not likely to be of any immediate threat.

The curious triangular eye ornaments make the puffin look as though it is shedding a tear. Note that the eye-ring itself is coral red, and that a small furrow runs from the base of the eye back towards the nape. **[LEFT]**

The fleshy yellow rosettes at the sides of the mouth are made up from wrinkly skin. They are merely ornamental in function, and are often targeted by puffins in combat. **[RIGHT]**

We shouldn't be complacent, though. There are two very serious threats currently faced by puffins, along with a number of other seabirds. Firstly, pollution of the marine environment could have very serious long-term effects, and secondly climate change could be potentially catastrophic. The main danger is that nobody can easily predict what the effects of either may be. If, for example, the puffin's favoured fish were forced to change their spawning grounds by warming and changing of currents, so that they were out of reach of puffin breeding colonies, this could cause a series of breeding failures and, eventually, a collapse in population. Any kind of interruption in the food chain, caused by an unforeseen alteration of the marine environment, could equally have damaging consequences.

For now, however, the puffin is still out there in reasonable numbers. It's a summer headline act on the slopes and cliffs of the North Atlantic, waiting to be seen and enjoyed. ▦

Puffins have longer legs than most other members of the auk family, and this enables them to move about much more easily on land. They frequently run about, something that a guillemot, for example, simply couldn't do without falling over. The tarsus (leg) is broader and rounder than you would normally expect for a swimming seabird; most seabirds have laterally flattened legs to reduce water resistance. This is a compromise between the puffin's need to swim and its need to dig burrows and walk. **[BELOW]**

In this picture you can see that the inner claw on each foot orientates to the side, rather than forward. This preserves it from the wear and tear of walking and landing, ensuring that it remains sharp for the important tasks of digging and fighting.   [LEFT]

With just the one definite groove on its bill, this is clearly an immature individual, probably two years old. It will be another two years before it will be old enough to breed.

[BELOW]

PUFFINS AT SEA

We are so used to seeing pictures of puffins at breeding colonies that it is easy to forget that they are actually seabirds, not land birds. However, one look at a typical puffin's diary puts the record straight. Between mid-August and early March, depending on latitude, these auks are exclusively seagoing, occurring well offshore and not within sight of land. That amounts to three-quarters of their year. Early on in the breeding season they don't come to land very often, either, but tend to gather in flocks close inshore, seemingly reluctant to get their feet dry. Only between April and August are they committed to their colonies, and during many of those days they aren't at the colony at all, but out at sea catching fish for their young.

Once out to sea puffins are in their element and not ours. They become difficult enough to find, let alone study. It is virtually impossible physically to follow one of these fast-flying birds out into the vastness of the open ocean and watch what it does on the surface. Out and sea and underwater is another realm where we can only go in our imagination. In recent years, though, we have been able to send technology with the puffins themselves to shed light on how they live in their true home, using GPS locators and depth indicators. Up until now there has been a particularly wide gap in our knowledge about a puffin's life during the winter, and only now it is being filled, albeit slowly.

〉

Puffins have evolved to live in the chilly waters of the North Atlantic. Their bodies are thickset and squat and covered with unusually dense plumage, twice as thick, for example, as that of a gull. The feathers are short and packed into large numbers of tracts, and there is copious down close to the skin to ensure that every part of the body is well wrapped up. The waterproofing qualities of the feathers, coated in preen oil, are obvious from the photo. This individual is a sub-species photographed in Svalbard. [RIGHT]

Puffins spend much of their life swimming buoyantly on the surface of the water, and they obtain all their food by diving underwater and pursuing fish. The sketchy information that had been gathered in the past showed that, at times, they were able to go as deep as 60m, as birds had drowned in fishing nets placed at that depth in Newfoundland. Recent studies show that that is very unusual, and when feeding puffins actually don't usually go very deep at all. Average depths recorded for birds feeding near their and also offshore in the non-breeding season came to the same value, about 4m, although puffins can, and commonly do go down to 35m. It does, of course, depend on where the fish happen to be.

What did enthral the researchers was not how deep their study birds went, but how frequently they dived. 'Little and often' was the rule, even over long periods of time. One early result clocked a puffin diving 194 times in 84 minutes, with an average dive lasting 28 seconds, with just 6 seconds recovery time before the next submergence. A parent with a chick in the nest made 442 dives in a little over 2 hours 30 minutes, spending 70 per cent of its time under the water. This frantic rate did relent a little, and an average count for four parents over the course of a day was 1148 dives (the birds did not hunt in darkness), during almost eight hours of active foraging. Of particular interest was the fact that, during this time, the parents actually consumed 90 per cent of all the fish they caught, and delivered only 10 per cent of them to the waiting chick.

⟩

Contrary to what you might expect for such a sociable bird that nests in claustrophobic concentrations, puffins are essentially solitary out at sea. They actually avoid other puffins and, even when several gather close together over shoals of fish, they have nothing to do with each other. In contrast, cormorants and gannets often feed in large numbers, which helps to confuse the fish and make them easier to catch. [LEFT]

It might seem odd for a waterbird to have a bath in the water, but many seabirds, not just puffins, frequently indulge in this kind of behaviour. It seems that regular soaking helps to keep the feathers in shape. **[RIGHT]**

This bird appears to be gathering preen-oil from a gland on its rump known as the uropygial gland. The gland secretes oil that is rich in fats, fatty acids and wax, and is good for keeping the plumage in prime condition. It also probably confers waterproofing properties on the feathers, although whether it is essential for this is unknown. The act of touching the gland stimulates it to release the oil. **[BELOW RIGHT]**

To maintain such a feeding rate puffins obviously need to be able to reach rich and easily accessible fishing grounds. They can obviously manage this almost all the time, because they are long-lived birds for whom 20 years is a normal lifespan. When a parent has delivered food to the nest and is ready to go fishing again, it flies out to sea without hesitation and evidently makes a beeline to a location where fish can be found. Presumably, early on each morning adults quickly discover hotspots offshore and other birds follow their fellow commuters out there, or perhaps they take a cue from non-breeding residents. Either way, puffins don't seem to wander aimlessly about looking for food. They obviously have a way of finding food supplies that we don't understand.

Obviously it goes wrong sometimes. Puffin colonies that have existed for many generations would only do so if there were abundant food supplies available at a realistic distance away. But fish populations, being biological systems, are not always predictable, and in some years, when they move or crash, the breeding of puffins suffers in kind. In some years, barely any birds at all are trying to nest, while the next year could see the colony bursting at the seams.

In the non-breeding season puffins only have themselves to look after (unlike, incidentally, razorbill and guillemot fathers that attend their young for some months after the latter have left the breeding ledges) and in theory can wander at will in search of a good feed. But where do they go? The winter distribution of puffins has for many years been a mystery, save for the fact that ornithologists assumed that the birds dispersed well away from the colonies, from where they were entirely absent between September and February. It was also obvious, from the paucity of birds seen on 'seawatches' (land-based observation of offshore birds) that, unlike razorbills and guillemots, puffins went out of sight of land.

Until recent telemetry studies, there was little to go on except for recoveries of dead puffins and from the occasional at sea reports. The latter showed that some birds travelled great distances; for example, British puffins have been recorded as far south as North Africa and Icelandic Puffins have been recorded across the Atlantic in Labrador. The at sea surveys from ships have found large concentrations in the north-western North Sea, off Scotland, but not many others. It is thought that puffins do not stay together in large, dense flocks, but disperse over the surface of the sea, making them extremely hard to detect.

Although in its infancy, the use of geolocators has begun to unravel the mystery. Studies of British birds have shown that different individuals may have radically different wintering strategies with, for example, some birds wintering in the North Sea and not moving far, and others navigating extreme distances into the North Atlantic, in at least one case almost half way across the Ocean. It also seems that an individual might choose different locations from one year to the next, or move around from month to month. As yet there is little or no information from elsewhere. But at the moment it does indeed seem that puffins may indeed wander where the fancy takes them.

⟩

The groups of birds that gather offshore in the early breeding season often contain birds that seem to be paired. It is known that members of a pair don't spend the winter together, so perhaps they meet up again on the sea rather than at the nesting burrow?

Gatherings such as this are usually only seen very close inshore by the colonies, and tend to be social in purpose. In fact, in early season puffins seem merely to flirt with the possibility of visiting their breeding burrows, keeping a shy distance out at sea.

It isn't unusual for puffins to spend a great deal of time loafing around on the sea; non-breeding birds near the colonies hardly seem to do anything else. Normally, these birds show a peak of feeding in the early morning and again in the evening, and they don't feed at night. So flocks can mill about on the surface simply floating about, sometimes washing and sometimes bickering. There is, if you like, a sense of drift. [RIGHT]

Another intriguing question is this: if puffins do travel long distances at sea in the wintertime, how do they do it? Do they fly – and if so, what clues in the water surface cause them to pause during their journey? Or do they swim, using their own power and the drifting of currents? The private life of the puffin at sea is still one that we know very little about. ▪

This stunning shot shows how far back the puffin's feet are set on its body, an adaptation to underwater swimming. By having the feet at the back, like the propeller of a boat, the large leg muscles fit neatly into the streamlining of the body. They are also in the best position to steer the bird when it is swimming. Note that, as is the case with most swimming birds, the feet are webbed. [BELOW]

Puffins on the water often hold their wings out like this and give them a few flaps. This can be interpreted as a flight-intention signal, although the bird doing it may not take off straight away. **[LEFT]**

In common with many seabirds, and most of the members of its family, the puffin is predominantly black above and white below. The dark colouration on the upper body makes it difficult for predators to see a swimming puffin from above, should they be flying over, and the pale ventral colouration makes it harder for predators, and possibly prey, to see the bird from below. Note, incidentally, that the puffin's underwings are black: its relatives the guillemot and razorbill both have extensive white on the underwing. **[LEFT]**

PUFFINS FLYING

Anybody who has seen a puffin flying will conclude fairly quickly that the air is not this bird's medium of choice. A birdwatcher spotting one flying over the sea offshore will simply see a blur of wings, and the impression gained is of a bird flapping almost desperately fast, as if it could stall at any moment. A visitor to the breeding grounds might see a puffin coming in to land, a manoeuvre that often ends with the bird's dignity compromised, as it bumps down or trips upon arrival. In the water, furthermore, gaining flight requires a run-up, with the feet pattering the surface, while landing sometimes involves a belly flop or a splash, especially in heavier seas. Puffins can fly, and sometimes quite fast, but they don't always look in complete control.

The puffin's wing is a compromise between flight in two mediums, in the air and in the water. As we've seen, in the water the wings need to be small and narrow, otherwise they will slow the bird down by friction, something that is unhelpful for a bird specialised in chasing food underwater. They are also heavier than the wings of land birds, with denser bones that reduce buoyancy in the water. On the other hand, puffins also need to reach their breeding burrows, meaning that they must be able to take off. The wings are conspicuously narrow, reducing their wing loading (the area of wing relative to the bird's weight), but the low surface area means that they stall at a relatively high speed. In consequence they must beat their wings fast to keep going, and landing can be something of a lottery.   ›

Wheeling around the snow covered cliffs of Hørnoya, the reflected light of which brightens up its underside, this bird is returning to shore for the first time after seven months at sea: no wonder it is looking carefully for a good place to land.   [RIGHT]

Nevertheless, a puffin's ability in the air shouldn't be underestimated. When fully underway it is thought that the wings beat at a rate of about 500 to 600 beats per minute, and the bird is capable of impressive speeds. Puffins can certainly fly at 100km per hour, and 50 to 60km per hour would be usual for a fishing trip in calm conditions. ▥

The fast flapping of a bird is sometimes though to create lift, but all it does is to thrust the bird forward, and the natural aerofoil shape of the wing provides the lift. On this image you can see the two main sets of flight feathers: the wing-tip is made up by the primary flight feathers, and lower down the wing main trailing edge is made up of the secondary feathers.   [LEFT]

This image demonstrates how small the puffin's wings are in relation to its broad and plump body. The ratio of a bird's weight to the surface area of its wings is known as 'wing loading'. The higher a bird's wing loading, the faster it needs to fly to prevent stalling.   [BELOW]

With its low stalling speed, a puffin
is greatly aided by strong winds
gusting around the cliff tops. In
such conditions, these birds might
actually look at ease in the
conditions. Indeed, they are
perfectly capable of keeping
motionless or gliding. Sometimes
they perform a display called a
'wheeling flight' over the breeding
site, describing wide circles or a
figure of eight.   **[RIGHT]**

This individual is arriving at its
burrow against a strong onshore
wind. The conditions meant that it
had to come in parallel to the cliffs
and fly against the wind in order to
manoeuvre its way in.   **[BELOW
RIGHT]**

This image demonstrates the power of the forces that the feathers have to withstand in flight – note how the wing tip bends upwards as this puffin prepares to come into land. [BELOW]

At times in flight the feet are spread.
They can help as stabilisers and
air-brakes.   **[LEFT]**

A puffin comes in to land. Note the
sharp claws, which help the bird to
grip when it touches down.   **[BELOW]**

Happily for hard-working parents, most feeding trips do not require long flights. Studies on Skomer, for example, have shown that 85 per cent of all trips were to waters less than 15km away, and often much less than that. On the other hand, depending on what resources are available, some flights can be very much longer, and a marathon of 137km has been recorded in Norway. [LEFT AND BELOW]

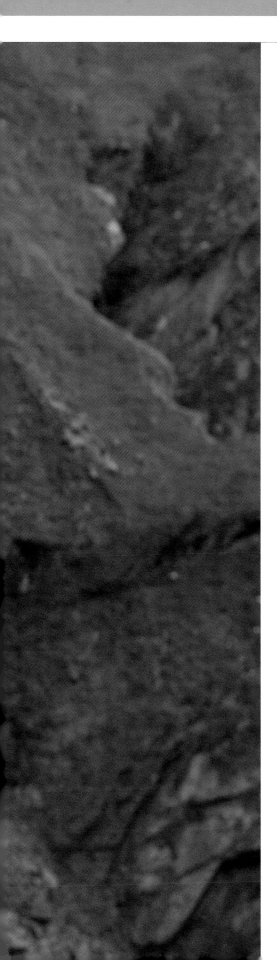

PUFFINS ON LAND

It's hard not to get the sense that puffins are reluctant birds on land. Their arrival at the colonies, for example, shows at best muted enthusiasm and at worst getting over an aversion. The departure later in the season is prompt and without ceremony; once the young have gone and there is no need for them to be there, the breeding birds are off in a hurry.

However, to land they must come if they are to breed successfully. The rigours of flight have meant that birds generally have never evolved the ability to nurture their young inside the body, so they must lay eggs and incubate them, something that cannot be done in the water. In common with most (although not all) seabirds, puffins nest as close to the sea as they possibly can, because this cuts down their journey time to and from their fishing grounds (it's like ensuring that you live near a supermarket).

However, not all places close to the sea are suitable. Being small seabirds, puffins are relatively defenceless on land, and when breeding they need to ensure that they are well out of the way of any ground predators, including cats, dogs, rats, red fox (*Vulpes vulpes*), Arctic fox (*V. lagopus*) and, of course, humans. You therefore don't find breeding puffins on fields or beaches, but on steep cliffs and offshore islands. Not all offshore islands are free of rats, for example, but those that are make secure breeding sites. ›

The puffin's flirtation with land is temporary. Only between April, at the very earliest, and August at the latest do these birds set foot on the breeding grounds. Once the young have fledged they return to the sea. **[LEFT]**

The puffin's apparent reluctance to come ashore to breed is amply demonstrated in the early weeks of the breeding season. They may turn up in the vicinity of a colony some weeks before they actually touch dry land. At first a few birds appear offshore in small groups. Numbers slowly build up only to drop again as some, if not all, suddenly disappear out to sea again, as if they had cold feet about what they were about to do. Eventually a few build up the courage to make short flights over the burrows, and only after this do the first birds finally make landfall. It is as if they are flirting with a mate about whom they are uncertain.

The bright orange legs and feet of the puffin are fitted with extremely sharp, curved claws. Although primarily used for digging, the claws are also useful for holding on to awkward or slippery surfaces, and can adminster a serious scratch to a rival puffin or, indeed, to a human researcher. [RIGHT]

Puffins only ever come ashore in the spring and summer. Outside this time they are occasionally 'wrecked' on beaches during autumn or winter storms, but rarely survive. [OPPOSITE]

Once they have finally landed for the first time at their burrows, you might expect puffins to spend much of their time ashore, but in fact this isn't the case. They do, of course, still have to feed themselves, and disappear for considerable periods of time out at the feeding grounds, often en masse. Indeed, whether they are out at sea or on land, by this time in the season they do most things together, in large groups, like human commuters. Studies have revealed quite a consistent daily pattern of attendance, with a definite build-up of numbers in the late afternoon and evening, and very few birds around in the early morning or in the middle of the day. Nobody has yet worked out what causes these preferences, although puffins can only feed during the day, so perhaps they spend most of the daylight hours doing this? Weather conditions also play a part in puffin attendance at the colony. In very stormy weather, and conversely in fine settled weather, not many puffins are about. Perhaps they prefer to come to land when there is enough, but not too much breeze to make their landings more comfortable? They don't seem to mind fog, though, as good puffin days often coincide with poor visibility. ›

The weather does affect the number of puffins present in a colony at any one time. If it is particularly stormy, birds will shun the colonies and remain out at sea. Conversely, they also stay away when it is still and sunny. [RIGHT]

One might expect a seabird to be immune to rain and indeed, adults within a colony may give the impression of revelling in it. However, puffins do nest in burrows and, in particularly wet years, these burrows may become unavailable through flooding or, even worse, actually cause death to the chicks by chilling or drowning. [BELOW]

There is an interesting correlation between colony attendance and the age of a given individual bird. The first birds to make touchdown are invariably experienced birds over four years old, and these make a bee-line to the burrow they have used before. The next age-class to arrive are the young adults and so on, until the youngsters, attending a colony for the first time, finally arrive, a month or more after the vanguard. In summer, therefore, a puffin colony is an all-age affair, which is by no means the case in all seabirds. In gannets, for example, the youngsters wander around well away from the colony during their first year, often spending the summer thousands to kilometres to the south of where they grew up. Not so puffins – the young birds might try out a new colony, or return to their own natal colony, but they will mix it in puffin society right from the start. ▨

The cliffs and slopes where puffins breed face the sea and are often very breezy and draughty places. Moderate winds actually help the puffins to fly to and from their burrows, but at times it can play havoc with the plumage. [OPPOSITE]

There are times when the puffin's ability on land begins to reach its limits! Webbed feet are not ideal for clinging on. [BELOW]

Puffins often attend their colonies early in the morning (as here) and in the evening. There is often a mid-day lull. **[LEFT]**

The rain doesn't exactly help when a puffin is digging a burrow. However, these are birds quite used to cold, wet conditions. No harm is done, and the plumage will be cleaned on this puffin's next visit to the sea. **[BELOW]**

Even on land the puffin is a seabird, as can be seen in this picture. Colonies are always within easy reach of fishing grounds. Note the bird on the far right. All it has to do is jump off, and it is airborne. **[OPPOSITE]**

Cliffs are typically favoured because they face the open sea and are windy, which helps puffins take off and land. **[TOP LEFT]**

Puffins are used to frost in the Arctic colonies. However, this image was taken in Skomer, Wales, early on an unseasonably cold May morning. **[BOTTOM LEFT]**

PUFFIN PLACES

Puffins would doubtless always be popular with people, given their colourful and singular appearance. However, it surely helps that some of the places where they breed are open maritime landscapes, scenic, bracing and spectacular, rather than buildings, or trees or farmland. To reproduce successfully the birds need sites close to the open sea and out of range of ground predators, such as rats and foxes, and hence they are drawn to isolated islands and precipitous cliff-tops, where they nest in high density along with other seabirds. To see them in the wild, therefore, is to embark on an adventure, often a boat trip, or a long walk or drive along remote tracks, away from civilisation. Puffin places are vivid and colourful, subject to the whims of violent weather; they instil awe and wonder, and a sense of wildness. Visits to such locations leave you with far more than just an encounter with a bird. ▨

An individual poses quietly during a long evening on Fair Isle in Shetland. Many puffin colonies are at high latitudes, where the nights are short and the days long. There is no doubt that, for diurnal feeders like puffins, the long days help them to locate enough food for their young. [RIGHT]

To the south of Hermaness, between the Shetland and Orkney Islands, Fair Isle holds a colony of at least 10,000 pairs, and probably many more. They are all but impossible to count on the sheer grassy slopes and isolated stacks. Behind this bird is Sheep Rock, one of the landmarks on this small, very isolated island. **[LEFT]**

Some birds on Fair Isle attempt to nest among these mounds of peat, the grass eaten away by rabbits. On this relatively flat ground the colony can sometimes look unusually crowded. **[BELOW]**

This view looks north from Hermaness towards the sea stack of Muckle Flugga, which does indeed mark the northernmost point of Britain. The soil at Hermaness is soft and thick and perfect for puffin burrows. Many nesting sites here are first excavated by rabbits, saving the resident birds a great deal of effort. **[ABOVE]**

Many of the Puffin's breeding areas are well within the Arctic Circle, so it is hardly surprising that they encounter snow from time to time. Here at Hørnoya in Norway, birds need to arrive early in order to claim their burrows, although that clearly won't be happening today. **[LEFT]**

Sitting or standing on the snowy and often very windy slopes of Hørnoya provides a new challenge to the birds on their return to land after a winter at sea. **[RIGHT]**

Many people have seen their first Puffins at Skomer, an easily accessible island in Wales with about 13,000 pairs. Nearby Skokholm has about 3500 pairs. Behind these puffins in May, the headland is painted by a streak of flowering bluebells (*Hyacinthoides non-scriptus*). **[BELOW]**

England does not have many puffin colonies, but situated 2.5km off the coast of Northumberland are the Farne Islands, with 27,000 pairs. In this image taken in May, you can see the landmark of Bamburgh Castle in the background. **[OPPOSITE]**

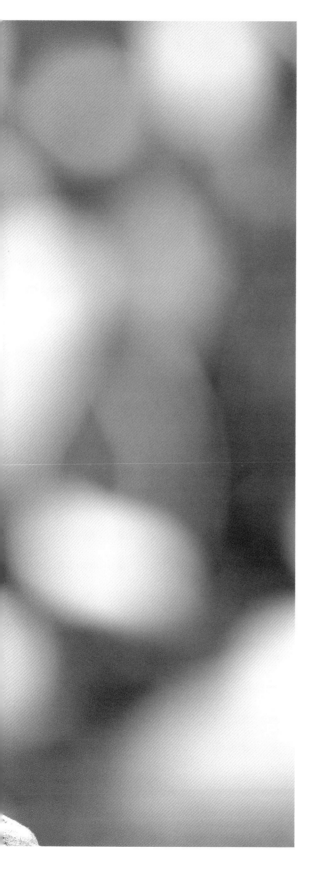

Puffins share their ledges with many other seabirds at Inner Farne, just as they do everywhere. In this image, the white bellies in the background belong to Guillemots. **[LEFT]**

In many places, puffins form the basis of a local industry. This seabird's enormous popularity ensures that, wherever colonies occur close enough to people, somebody with a boat will organise trips to see them. This is at Seahouses, Northumberland, the gateway to the Farnes. **[ABOVE]**

It should be borne in mind that seabird colonies persist over many generations, and the Hermaness colony, for example, has been protected as a reserve for nature since 1831. Before then generations of Shetlanders will have hunted the seabirds, perhaps from Mesololithic times 6,000 years ago. It isn't fanciful to suggest, therefore, that puffins have probably sat upon this very outcrop for hundreds of years.   [OPPOSITE]

On Skomer, off the Pembrokeshire coast, puffins often nest among flowering Sea Campion (*Silene uniflora*). Legend has it that picking Sea Campion brings bad luck, which is certainly true if you muck around the sheer cliffs where it grows.   [BELOW]

Some places, such as this outcrop at Hermaness once again, seem to have an almost magnetic attraction to puffins and will almost always seem to have a bird or two resting there. Maybe it's just a particularly good view from there! **[ABOVE]**

As for most puffin places, you need a head for heights to work at Hermaness. This image encapsulates everything a puffin needs for breeding: the sea in close proximity, precipitous cliffs or slopes to deter would-be predators, and soft soil in which to dig the burrow. **[RIGHT]**

Only the most blinkered devotee of puffins would be able to overlook the profusion of flowers that can be found around cliff-top colonies, many of which reach their peak of flowering at the time the birds are breeding. Here at Great Saltee Island, off the coast of County Wexford in the Republic of Ireland, the blooms of Thrift (*Armeria maritima*) can turn whole cliff-tops pink. **[OPPOSITE]**

Here on Látrabjarg in Iceland, a puffin has evidently made itself extremely comfortable on a clump of Common Scurvy-grass (*Cochlearia officinalis*). This plant usually grows by the sea, and was indeed eaten by sailors to cure scurvy. **[ABOVE]**

At Sumburgh Head on Shetland, a puffin is surrounded by wildflowers, this time the pink of thrift and the white, daisy-like flowers of Sea Mayweed (*Tripleurospermum maritimum*), shining in the last rays of the sun during one of Shetland's long evenings. **[LEFT]**

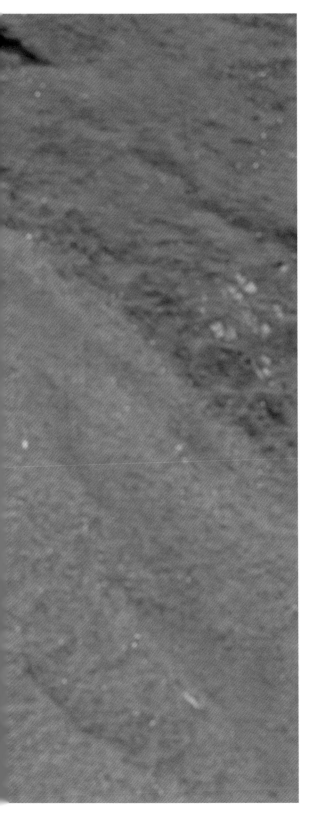

Hermaness holds one of Britain's largest puffin
colonies, with about 25,000 pairs. The birds here nest
along 3-4km of cliff, grassy slope and scree.
Hermaness is a promontory that juts out northwards
from Unst, the northernmost island of the Shetlands and
thus the most northerly point of the British Isles. The
waters around Hermaness are very productive, and in
total the site hosts well over 100,000 pairs of seabirds
of many different species. **[LEFT]**

A typical puffin landscape. These seabirds spend the
summer in out of the way places, often on tall cliff tops
overlooking the sea, and on isolated islands. Find a
puffin and it will invariably be in a spectacular setting.
[ABOVE]

This puffin was photographed by the light of the midnight sun at the cliffs Látrabjarg in Iceland, Europe's most westerly point. Iceland is one of the largest and most important breeding grounds for Atlantic Puffins with an estimated three million pairs breeding there each year. **[RIGHT]**

Early socialising on the snowy cliffs of Hørnoya offers much more than just a lesson in keeping your balance. **[BELOW]**

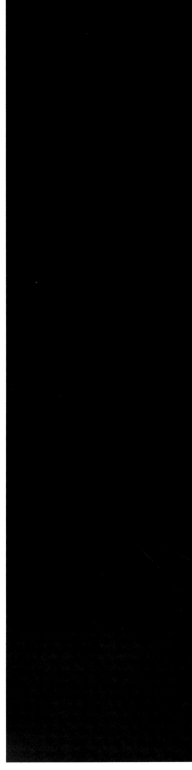

GROOMING AND RESTING

Anybody who has observed puffins at close quarters in their colonies will tell you that their subjects spend a surprising amount of time not doing very much. This applies especially to non-breeding birds, which always make up a significant proportion of individuals attending the breeding areas. The truth is, they don't have much to do: no pair-bonding, no fighting, no digging and no looking after of eggs and young, so they often simply sit around. However, it isn't only these feckless youths that give an appearance of idleness: paired up birds and parents don't hesitate to take many time-outs during the day. They don't spend all their time catching food, but instead do so in somewhat frenetic bouts, and at times they need to rest and sleep.

All birds sleep, and puffins are no exception. The amount they sleep varies, and nobody knows much about it: how much they need, or whether they ever get over-tired or have sleep deficits. Puffins certainly are largely inactive at night (or at least, they don't feed), and their rests during the day suggest that the breeding season is a tiring time. But the area of sleep is a mysterious one for scientists.

The posture of this puffin doesn't suggest that it is in deep sleep, although its eyes are closed. Most sleeping birds, including puffins, don't keep their eyes closed for very long, a few minutes at the most. **[RIGHT]**

Plumage care is one of those behaviours that is not of great interest to birdwatchers, but is of urgent concern to the birds themselves. Keeping the plumage in prime condition is absolutely essential, and can be highly time-consuming, especially during the moult. The main type of care is called preening, in which birds use their bills to re-shape their feathers – feathers are made of hundreds of small branches that 'zip' together, something that is performed relatively easily. The feathers also need to be kept in their tracts, which can be performed by preening, or even simply by shaking the plumage. Feathers don't simply sit on the skin, but are under muscular control and can be raised and lowered. Birds also use their feet to scratch hard to get at areas, and regularly bathe in the water, probably because it makes it easier to preen the feathers. Some birds sunbathe as well, but this has not been observed in puffins.

Birds preen assiduously, at regular intervals throughout every day of their lives. When arriving at their colonies from the sea, it's often the very first thing they do. This individual is working on one of its primary feathers (those that make up the wing-tip). But clamping its mandibles on the feather, it can 'zip up' the small branches that make up the feather structure. [RIGHT]

Feathers, especially the contour feathers of the wings and tail, get out of shape very easily, just through the usual wear and tear of everyday life. Not a day passes when a bird will not attend to them. [BELOW]

Preening, scratching and bathing all come under the umbrella term of 'comfort behaviour'. For the purposes of this section, excretion can do, too. And of course puffins do this as does any other species of bird. Puffins don't drink; all their fluid requirements come from food. They don't urinate as such, but regularly expel a combination of uric acid, urine, mucus and other faeces expelled as dropping. Any excess salt is removed through glands in the eye orbits in solution, making it look as though a bird is 'crying'. ▦

A shower of rain can be a bonus for a puffin, since water is known to help the birds when they preen, possibly by making feathers easier to reassemble. It's difficult not to come to the conclusion that puffins relish a good soaking. **[LEFT]**

Puffins dislike being dirty every bit as much as we dislike being unclean, or wearing worn or dishevelled clothing. For a puffin, however, being clean is essential for its survival. As well as dirt, preening can remove feather lice and other parasites. **[BELOW]**

Preening and shaking go together, and both can make puffins look like contortionists. It's particularly hard to reach the crown feathers. [OPPOSITE]

The dust is flying from this wing-flapping bird. [LEFT]

Better out than in! Remarkably, scientists have worked out the lag time between ingestion of certain fish and their final ejection. For a sand-eel it's 4.9 hours, for a whiting it's 5.7 hours, and the time from ingestion to sprat splat is 6.1 hours. The sprat is the largest of these fish. [BELOW LEFT]

Attention to detail is important to a preening bird. Ideally, not a single feather should be out of shape or out of place. **[RIGHT]**

With a little effort, a puffin can reach almost any feather on its body, with the exception of the side of its head. This bird is possibly aiming for the uropygial gland near the rump, which produces preen oil. Preen oil contains a mixture of healthy substances, including antibiotics and, possibly, components that react with sunlight to form Vitamin D. **[OPPOSITE]**

When a puffin is taking proper rest, as opposed just to loafing, it rests its head on the scapular feathers on its back. The eyes may be open or closed.   [RIGHT]

Feathers come in various types, from the stiff contour feathers than make up the wing feathers, through to the down that covers the skin. All types need to be attended to.   [BELOW]

This puffin is clearly unruffled by the rain shower, taking an opportunity for some rest. Feathers are generally waterproof. **[LEFT]**

Studies have shown that, at least in some species of birds, males sleep less deeply than females. But on the other hand, when birds rest in groups, their rate of 'peeking' decreases as they feel safer. The same happens when they rest in a safe place, out of reach of predators. **[BELOW]**

COLONY LIFE

Puffins are highly colonial, breeding together in assemblages ranging from a handful of pairs to tens of thousands. A few colonies, on St Kilda in the Western Isles, for example, encompass more than 100,000 burrows. (Puffins are, as you can imagine, exceedingly difficult to count – packed together on isolated and inaccessible islands, and hiding down burrows to boot). Clearly a puffin colony is a cramped, competitive and highly sociable place. Furthermore, since puffins are long-lived birds, the colony is cramped and competitive with birds that probably know each other only too well from previous years. On the whole, puffins return to the same burrow each year and pairs meet up at or around the colony, so it's reasonable to expect that rivals and neighbours find themselves in close proximity year on year as well. Scraps with those next door must be an important component in a puffin's social life.

In common with other birds, puffins have a wide range of visual displays to communicate their motivations to those in the near vicinity. Without these the whole colony would descend into anarchy. There are displays to claim burrows, to express aggression, to communicate appeasement and to express flight intention, as well as a whole range of sexual displays. Indeed, one of the joys of visiting a puffin colony is precisely to watch this sort of thing going on. The activity in seabird colonies is like a soap opera, only their dramas are, of course, real. If you witness an intruder making an attempt to steal a burrow, for example, the consequences of the action are real and serious.

Puffins are extremely sociable and actively seek out engagement with other birds. Living in such close proximity, it is essential that they communicate with a series of gestures and displays that other birds instantly understand. **[LEFT]**

One aspect of puffin colony life is highly unusual compared to most seabirds and indeed, to colonial birds anywhere. If you visit a puffin colony you won't necessarily notice anything strange at first. But if you were to jump from puffins to a cliff-ledge crammed with guillemots or better still, kittiwakes, you might suddenly realise something: puffins are remarkably quiet. There is none of the ear-splitting squealing of kittiwakes, nor the groaning or razorbills or the witches coven-like cackling of fulmars. Indeed, you can sometimes visit puffin colonies and hear nothing at all.

As it happens, puffins do call, but they do it sparingly. The calls are guttural and low pitched (some, maybe even most, are too low for the range of the human ear) and, with the exception of the odd grunt above ground, are made by pairs in their burrows, sometimes when one member is greeting a mate 'at the door', so to speak. The most prolonged calls are heard towards the beginning of the breeding season, and sometimes many pairs all call at once together. And in view of the fact that low-pitched calls will travel well through the earth, it has been suggested that calling might actually help to prevent puffins digging through the walls of their neighbours. If this were so, it would be a highly unusual function for a bird vocalisation.

Puffins live in each other's pockets during the breeding season. In a dense colony, neighbours are never far away. And since puffins are long-lived birds, the same individuals may well breed side-by-side year upon year, which presumably can be a blessing or a curse! **[LEFT]**

Dense concentrations such as these are found on the periphery of the actual colonies. Most of the birds' attending these so-called 'clubs' are young birds in their second or third years of life, still too young to breed. **[RIGHT]**

If you do visit a puffin colony from mid-season onwards, you might find yourself watching something quite different to what you were expecting. That's because a large number of non-breeding birds typically occupy the periphery of the colonies. They might stand or sit close to a burrow, but more often they have a section of cliff, or some rocky outpost to themselves. There may be tens or hundreds of birds seen there, often simply sitting or loafing about, but sometimes displaying in a half-hearted manner, leading you suspect, wrongly, that they are breeding birds. These communities of non-breeding birds are known, somewhat amusingly, as 'clubs'. Most of the birds making up numbers are two to three year olds. The function of clubs is presumably for young birds to learn social mores in a colony situation, although it is not entirely impossible that first meeting between future mates could occur here. If you visit a colony in a temperate location in late June and July, most of the birds you see will probably be these non-breeders.

Not surprisingly, most of the really interesting and important behaviour in a puffin colony occurs in the early stages of breeding. Upon first arrival birds must lay claim to a territory, protect that territory, dig or refurbish a burrow and ensure that they have ground fit for an egg or a chick – and that is if they are already paired (see also next section). Those few days and weeks are momentous for the puffins, and truly exciting to witness.

It isn't always obvious why puffins gape. In some ways it is a common language communicating aggressive tendencies. Note how the middle bird has been ringed as part of a study. **[RIGHT]**

There are one or two puffin displays that aren't illustrated here. Two are associated with protecting the burrow when a rival lands close by. One is known as a 'spot stomp', in which the individual raises and lowers alternate feet several times in succession, and action not dissimilar to our own lifting of our heels when we are standing up and nervous. In the puffin's case, attention is inevitably drawn to the colourful orange feet. The second display is known as the 'Pelican walk', and involves a bird walking around the entrance to its burrow in a highly stylised manner, with body vertically erect, plumage ruffled, bill pointing downwards. The feet are lifted upwards alternately, slowly and with exaggerated movement, as if the bird was marching, or doing some exercises. It is in response to another adult passing by. Puffins moving past other burrows often walk quickly and with their heads down and body horizontal, an appeasing 'low-profile walk'.

This individual on Inner Farne quite literally has its head down as it manoeuvres through the colony on a Low-profile walk, its posture appeasing. **[LEFT]**

Gaping is performed in many situations and at different intensities. This bird is not displaying its tongue, suggesting perhaps that this is low intensity threat. **[BELOW]**

Watch out, too, for 'Wheeling flights', in which groups of birds take off en masse and flying back and forth over parts of the colony, flying downwind over the sea and into the wind towards the land, often performing a figure of eight at the same level above the sea. Individuals taking part in these movements have been shown to be mainly non-breeding birds. ▣

Here a 'club member' performs a gaping display. The threat seems to be directed more at the photographer than to any other puffin, but performing these sorts of displays is vital to the maturing process. Gaping is a common display seen almost through the breeding season. It is essentially a threat display and is used by both sexes. It doesn't need much motivation, as seen here. The image was taken on Inner Farne. **[LEFT]**

Clubs may not always be lively places. However, individuals will indulge in bickering and half-hearted displays. This is puffin youth! **[BELOW]**

Birds that have just landed often perform a special posture with their wings open and body tilted forward. This is intended as appeasement, and is usually seen when a bird lands among a group. [LEFT]

A puffin raises its wings upon landing, a message that says, 'I'm here'. In the snow, lifting the wings can be somewhat perilous. [ABOVE]

Gaping is a common display seen almost through the breeding season. It is essentially a threat display and is used by both sexes. It doesn't need much motivation, as seen here. The image was taken in the Shetlands late in the evening. [OVERLEAF]

This might look like a conspiratorial tête-a-tête, but the bird on the right is being decidedly threatening. Note how the tail is cocked upwards and the head bowed down – the latter is something of a universal sign of aggression, seen in many species of bird. **[ABOVE]**

Fights between males are frequent in the early breeding season, especially if breeding burrows are in short supply. Normally the contest is short and sharp but sometimes, as here, birds get a good hold of each other and neither is willing to let go. Note how the wings are held stiffly out so that each bird can keep its balance – and balance is the key here, because if a bird falls over it will be at a severe disadvantage. When fights break out they invariably attract a fascinated audience of other puffins. These birds are on a footpath on Skomer Island, Pembrokeshire, and it's obvious that they are quite oblivious to the photographer. **[RIGHT]**

Once again, an onlooker intrudes into billing between two birds. Sometimes the spectator actually interferes with the courtship display, leading to an aggressive riposte. [LEFT]

Confrontations are routine in the rough and tumble of the crowded colony. Both these birds have a dirty bill from digging their burrows. [ABOVE]

This bird is quite clearly close to its burrow and, judging by the appearance of the bills of both birds, is threatening a near neighbour. Contests of this kind are common, especially during stage of digging or refurbishing the burrow. Normally the neighbour will simply gape back in kind, and normal service will be resumed. **[LEFT]**

Puffins engage in a lot of fights, which are invariably over burrows, of which demands always outstrips supply, and are most intense during the brief egg-laying period. Combat usually takes place in the evening, when more individuals than usual are present at the colony, and these are one of the few moments when these birds make a lot of noise – a loud growling. Tussles usually look worse than they really are, but birds have been known to roll down steep slopes in a ball and even disappear over cliff edges, like a couple of stuntmen on a movie. Even if not so fierce, fights often take the combatants well away from the burrows they are trying to defend. The tussles cease when one bird breaks away and flees. **[ABOVE]**

The individual on the right is threatening its neighbour. In full gaping threat display it stoops down and opens its bill, ruffles its feathers and, like a human schoolchild might do, shows off its tongue. **[RIGHT]**

PAIR FORMATION

Considering that the puffin is such a well-known and extensively studied bird, surprisingly little is known about how males and females actually pair up – particularly when they first get started. What is known is that, once they have bred, males and females meet up again at the burrow year after year, in what could be called a lifelong pair bond. However, as in the case of a number of seabirds, they are not thought to associate at all once they have left the breeding cliffs for the winter.

Divorce does occur among puffins, though. In fact, researchers have been able to calculate the average divorce rate at a number of colonies. On Skomer, Wales, it was found to be 7 to 8 per cent, on the Isle of May in Scotland 7 per cent and on Gull Island, Newfoundland it was 9 per cent. These, incidentally, refer to genuine divorces, when both members of a pair survived the winter and returned to their usual burrow. The cause of divorce seems mainly to be a rival male taking over a burrow and acquiring the female at the same time. In contrast to the situation in some other seabirds, such as albatrosses, divorces don't seem to be so traumatic that they affect an individual's breeding success. It seems puffins just get on with life and often successfully breed straight away with their new partner. ⟩

Puffins usually pair for life. The relationship is unusual because it appears to be entirely seasonal. Members of the pair are not thought to associate at all when they go out to sea in the winter (and individuals have been tracked in winter far apart), but at the colony they find each other again and will re-form their partnership.
[RIGHT]

Divorces do occur in puffin society, but they are not very common. Less than 10 per cent of pairs break up overall in a given year. **[LEFT]**

Billing is the most conspicuous courtship behaviour performed between puffins. It is almost always performed by paired birds, but occasionally the odd individual will cheekily play up to another bird's mate. **[ABOVE]**

Interestingly, these intensive studies show that divorced birds don't burn their bridges very far. On average, a divorced male is only displaced 1.7m, which is often just to a neighbouring burrow, whereas females move away an average of only 3.3m. Bearing in mind that these long-lived birds tend to get to know their neighbours well, it seems divorced birds simply burrow-hop into familiar territory with a bird they have already seen at a not very great distance. Bereaved birds move away much further, males on average 4m and females 10m – sometimes to a completely different part of the colony.

At times the billing display can get quite intense and the partners clash their bills together, making a loud noise that attracts the attention of bystanders. Note how the bird on the left has ruffled its feathers, which is part of the display.   [LEFT]

Not every attempt at billing is quite as smooth as it should be. It is difficult not to see some kind of discomfort, or even surprise, in the posture adopted by the bird on the right.   [OPPOSITE]

Going back to the original pair formation, the evidence suggest that the bulk of this takes place, not at a nesting burrow or potential digging site, but on the water – and this is why it is not yet fully described by people who study puffins. Certainly plenty of courtship goes on just offshore, close to the colony, but as yet nobody can be sure how 'serious' it is. A male was observed in Norway making its way through the possibilities at one of those offshore rafts. It displayed to seven different females in the space of twenty minutes, which suggests not very serious at all. On the other hand, during five hours of observation at a North American colony, 56 birds were observed to solicit 100 females, with the result that mounting occurred 34 times, 9 times with apparent success. This took place on the water, and it is pretty certain that almost all copulations occur when the birds are at sea. Birds will try on land, but apparently they are rarely successful in achieving what they are intending to do, perhaps because they only do this in fits of sexual fervour, and are not properly prepared. Incidentally, when they mount on the water, males remain in situ for an average of 24.5 seconds.

Nevertheless, in the early season the visitor to a puffin colony will see ample signs of affection. Birds will be indulging in a number of displays that leave no room for doubt about to whom they are paired. The most obvious display, and by far the easiest to see, is called 'billing', in which the pair touch bills and often seem to clash them. It can be seen at any time that puffins are at their burrows, and is performed both on land and sea.

There are other displays that are associated exclusively with pair formation and appreciation. One is 'head jerking', basically a prelude to copulation, and another is 'head bowing', a seldom seen manoeuvre in which the bird bows its head so much that it is almost touching the ground (or the sea). Only the male performs the first one, but both sexes may practise bowing. ▪

This puffin is performing the head jerking display. It is done by a male bird close to a female, and is typically an invitation for copulation. The male jerks its head upwards at the rate of once every second. The display may continue for as long as 10 minutes, and can also be seen when birds are displaying on the water. [BELOW LEFT]

Another bird head jerking, this time with the head at its highest point. This display has a vocal accompaniment, a quiet grunt. [LEFT]

Regardless of the weather, the need to make a mark through head-raising trumps everything. Besides, Puffins are relatively immune to the cold. [BELOW]

This image shows an interesting variation on billing. These individuals on Hermaness, Shetland, were picking midges off each other's plumage.   [RIGHT]

The muzzling of the partner's plumage is not preening, in the sense of keeping the feathers intact. Instead it is simply a courtship procedure that usually enhances and confirms the pair bond.   [BELOW]

Billing begins when one partner makes an approach to the other, keeping its head low. As it gets near, it swings its head from side to side and usually begins to preen the other bird's head or breast plumage.   [OPPOSITE]

The physical action of billing means that the birds embark on a sort of dance on the spot. Both birds here have their wings held open to avoid overbalancing.
[LEFT]

This is an unusual sight, because puffins almost invariably copulate on the sea, unlike most other seabirds. In this photograph the female is lying on the grass while the male flaps its wings vigorously to keep balance. Land matings such as this one usually occur in a moment of reckless early-season passion, and they rarely lead to fertilisation. And, let's face it, it doesn't look as though this couple has quite got it right! [**BELOW TOP**]

Whether it has 'worked' or not, copulation is an important part of keeping the pair bond strong. Here the male continues to flap its wings and jerks its head upwards.   [**BELOW BOTTOM**]

A pair indulges in a private bout of courtship. The
display is often contagious and routinely induces
excitement in the relevant corner of the colony. It's not
unusual for several other birds nearby to run down to
the displaying birds and watch or interfere – but
evidently not this time! In fact this image is from late
in the season, when breeding fervour has generally
died down. **[ABOVE]**

THE BREEDING BURROW

While many of their auk relatives, such as Razorbills and Guillemots, lay their eggs out in the open on precipitous cliff faces, where one false move can spell disaster, puffins lay theirs in burrows up to 2m long. While this strategy affords protection and shelter, it does mean that the burrow has to be dug, which has the disadvantage of being hard work and, potentially, making the birds vulnerable. Some pairs of puffins simply reoccupy burrows they have dug before – and indeed, experienced pairs arrive early and fight hard to keep their sites. A few birds in areas such as Skomer might strike lucky and simply refurbish a rabbit burrow. But for many, there's no avoiding moving what for a comparatively small bird is a great deal of earth.

The burrow is no mean construction. It takes about two weeks to build, although the birds don't spend any more than a few hours on any given a day actually digging, but equally important is its defence. Fights routinely break out between the birds' first arrival on land and the period when the first egg is laid, and the ownership does change hands, especially if a given burrow is occupied by a newly formed pair which find themselves ousted by more experienced birds. Very occasionally birds have been found using two branches of the same tunnel.

As mentioned above, tunnels are up to 2m long in normal circumstances, with most between 70cm and 110cm. In a few cases, such as Russia's Ainov Islands, the burrows can be interconnected like an underground labyrinth, each pair having its own side-branch. Some such tunnels have been measured at 15m, and they give a whole new dimension to puffin behaviour – what is effectively a seabird walking deep underground, digging in complete darkness and carrying fish into the abyss when its young is growing. Would puffins in such locations compete over burrows, and if they did, would they do so underground? ⟩

This image shows a typical burrow entrance. Note how it is on a slope. Slopes enable a higher density of burrows to form without the danger of the soil collapsing upon them. A typical burrow will have a diameter of 13–18cm and will be dug 1–2 metres into the ground. **[RIGHT]**

Every colony is different and not all colonies have suitable soil for making burrows. Sometimes there is no earth at all, and the birds must instead choose gaps between rocks on cliff faces. If a colony is particularly well populated, pairs may requisition unusual sites, such as in a wall or other artefact. Some birds in Arctic colonies have the difficult task of burrowing through permafrost if they are not using crevices instead; they often succeed. And, of course, even within a colony not all sites are ideal, even those places with puffin-friendly maritime peaty soils. That's because slopes are very much more suitable than flat ground; a burrow on a slope goes naturally deep even when dug horizontally, affording it a certain stability. On flat ground a burrow not only tends to have a flimsier roof, but that roof is far more likely to be compromised by, for example, by livestock treading over it. On flat ground puffins tend to seek out a slight hump that they can dig into first, but the resulting burrow invariably lasts for a limited number of seasons. No wonder, then, with so many variable aspects to a burrow, that pairs fight vigorously over the most suitable sites.

This image records a bird visiting its partner. The member of the pair actually doing the work is underground. From this picture you can see that this burrow is very much under construction. The flying soil results from the bird kicking the loose soil backwards out of the nest. It tends to flick it with one foot at a time, rather than resting on its belly and using both. [OPPOSITE]

Although the puffins' core breeding habitat is soft soil on cliffs and islands, this simply isn't always available, especially in the Arctic where birds would have to dig through the permafrost. Large populations, therefore, just use holes among boulders and rocks, or scree slopes. This puffin was photographed at Látrabjarg in Iceland. [BELOW]

The burrow itself is quite narrow, 13 to 15cm in diameter, not enough for birds easily to pass each other. It ends in a nest chamber where the egg is laid, usually upon a small pile of nest material. Interestingly the puffin puts quite a lot of effort into nest sanitation. The youngster, when it hatches, does not simply defecate where it sits, fouling the burrow. Instead the parents fit their burrow with a side-chamber, where the chick assiduously goes to relieve itself. For their part, the parents don't defecate inside the burrow, but go just outside to do so, albeit virtually at their own doorstep. Interestingly, they seem to defecate on land a lot, where you might expect them to do so usually at sea. ›

Both sexes play a part in burrow construction, although it seems that the male often takes slightly more of the brunt of hard labour. There isn't room for both birds to work at the same time, so perhaps this photograph records one bird relieving the other. **[TOP LEFT]**

Flat ground presents a problem for puffins, because of the danger of roof collapse. While burrows on slopes go naturally deep with an even floor and are relatively stable, this doesn't apply to burrows on flat surfaces. The birds cannot dig straight down, because they would be unable to kick the earth away effectively enough, and any chick caught in a hole would be unusually vulnerable to heavy rainfall. Flat ground colonies often don't last long. **[BOTTOM LEFT]**

The amount of material brought in to line the nest varies considerably, from none at all though to luxurious. So, while some eggs simply rest on the soil, others are effectively on a mattress. Where there are a lot of stones or boulders in the floor or walls, this material can act as a buffer against the egg getting chipped, although there isn't a direct correlation between the substrate and the amount of nest lining. Most nest material consists of dried vegetation that is plucked from dry land near the burrow, but birds sometimes bring in seaweed and during the season almost any kind of debris acts as a magnet for a puffin's curiosity. It is most entertaining to watch puffins playing with grass, twigs, fishing line, leaves and flowers. Most such playing seems little more than fiddling, and when a puffin is taking nest lining seriously, it simply flies into the burrow with a beak-full without any fuss.

It isn't only in the permafrost zone that puffins use rocky crevices. These birds are at Hermaness, in Shetland. [BELOW]

No puffin was harmed in the capturing of this image. The individual did manage to extricate itself. **[LEFT]**

Novel items are a source of curiosity and, evidently, worth to puffins. These birds are having a tug of war over a blue piece of string and, as you can see, the bird on the left is in a puffed up, threatening posture. **[BELOW]**

Later in the season, when the young have hatched, puffins sometimes take more freshly plucked vegetation into the burrow. In some species of birds, fresh green material is known to act as a sort of disinfectant for the nest, and this could perhaps be the case for the puffin, too.

One essential structural quality of any burrow is that it must be dark in the egg chamber. Young puffins are born with an aversion to light, and therefore the tunnel must be long enough to ensure that their need for dark is met; alternatively it can bend away from light. Some puffins have tunnels that are so short or ill constructed that a visitor can see the incubating bird; such constructions are doomed to failure. ▥

Once puffins have dug a burrow, they usually line the nest chamber with some kind of material, typically nearby vegetation. The amount of material varies according to the available supply. Occasionally, some individuals bring in seaweed. [RIGHT]

This is serious collecting of material. The bird yanks off grass or flowers by their roots, and will make a purposeful beeline for the burrow. [BELOW]

Digging can be particularly messy in the rain. Note how the inner claw curves to the right, rather than down as the other claws do. This is an adaptation to keep it out of the way, so that it will remain sharp for digging and fighting purposes. The other two claws face forwards. **[ABOVE]**

It's hard to say whether this is an interfering mate or a neighbour with designs on the nest material. Both are entirely possible. **[RIGHT]**

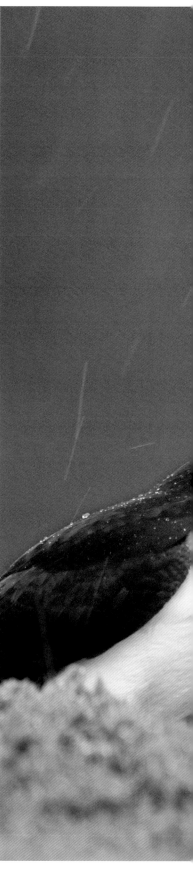

Any bill that can hold wriggling fish should be ideal for carrying nest material. The way puffins often 'play' with plant matter and other foreign material in their bill suggests that it is of some benefit to them.   [**LEFT**]

For these sociable birds, not even the hard work of digging a burrow will prevent moments of neighbourly interaction, although this one seems to be unusually peaceful. In this image, it's hard not to evoke the idea of three grubby workmen taking a tea break.   [**BELOW**]

Puffins are inveterate fiddlers, and any kind of material brought in mischievously by the photographer will arouse the birds' curiosity. This non-breeding bird is investigating a piece of string towards the end of the season. Although it looks strange to us as nest material, a piece of twine is fine for a puffin. Nesting chambers routinely house manmade items such as these. **[BELOW AND RIGHT]**

A bout of fiddling might well have produced a world first – this bird actually managed to tie a knot in the string! **[BOTTOM]**

This hardly constitutes much of an effort towards nest building, but a close look at the bill of this bird shows that it is still too young to breed. For youngsters, incipient material collection is just another type of practice for the real thing in later seasons. **[LEFT]**

Puffins use a wide variety of items for their nest lining, but mostly, and prosaically, it is just dried grass. **[LEFT]**

Later on in the breeding season, adults often bring fresh plant material into the nest chamber. This individual has a bill full of Sea Campion. Who knows – perhaps this plant has some use for freshening up or even disinfecting the burrow? **[BELOW TOP]**

The urge to collect nest-material can have unforeseen consequences. This bird is seemingly finding this dried out corpse of a Manx Shearwater (*Puffinus puffinus*) fascinating. (Incidentally, the odd scientific name of the shearwater is from the same root as the puffin's English name. The root is the word 'puff', meaning swollen or fat, and was originally used for young shearwaters harvested from their burrows for food. Puffin, then, is the borrowed name.) **[BELOW BOTTOM]**

EGGS, CHICKS AND FEEDING THE YOUNG

Sometime between late April and late May each year, puffins begin to lay their eggs. Not only does the date vary between years, but colonies in the north of the puffin's range begin later than those in the south. In common with other colonial-nesting seabirds, most pairs follow a reasonably synchronised routine, so that everywhere about the colony the majority of chicks reach the same stage at the same time. There are always exceptions to this, of course, with a few pairs making up for lost time for personal reasons: late arrival at the colony, not being in the correct physical shape, re-laying when a first egg is lost, and so on. The occasional egg is laid as late as July. Within a given colony there also tend to be clusters of super-puffins, birds that are ahead of the rest, and these pairs are consistently more successful breeders than their colleagues elsewhere in the colony.

Each female puffin lays a single egg. Interestingly, the puffin has two brood patches (layers of bare skin used to incubate), so presumably sometime in the distant past the species used to lay two. The egg itself is whitish, typical for a bird that breeds in a cavity – it's easier for the parents to see it. It is quite large, about 6cm long, and is more pointed at one end than the other, although much more rounded than the egg of a guillemot, a relative of the puffin that breeds on sheer, open cliff ledges. In that species the egg has a narrow turning circle that may ensure that is doesn't easily roll of the ledge. The weight of the puffin's egg is around 15–16 per cent of the body mass of the female.

Once it has been laid, the egg is incubated in turn by both parents for a variable period, typically within the range of 39 to 43 days. Incubation shifts vary greatly, but most last just over a day, ›

This is a highly unusual sight, an adult incubating its egg in sight of the su rface. Such shallow burrows rarely produce a successful outcome for the breeding pair. **[RIGHT]**

with an average of 32.5 hours. During this time the parent doesn't grimly sit on the egg without daring to move; inside the burrow the egg is well protected, and the bird in change frequently takes breaks, sometimes even popping out to sea for a short while, on average for 7.4 minutes (but up to 22). When it is incubating the adult places the egg under one of its brood patches and holds it in place with its wing.

The puffin chick is quite well developed at hatching, especially compared to a songbird such as a Blue Tit, which is naked, blind and completely helpless. Instead a puffin is covered with long, soft, black down, except on the chest and underparts where the colour is whitish, reflecting the contrast it will show throughout its life. The chick's eyes open very quickly after hatching and, once it has dried, the down gives it a reasonable degree of protection, so much so that it can survive on its own in the absence of its parents. However, in normal circumstances the chick is brooded for the first six or seven days of its life, when the parent will sit over it much as it would incubate an egg. After this the chick is able to regulate its own body temperature, and no longer needs such attention. From now until fledging it will remain in the burrow, waiting for its parents to bring in supplies of fish.

This individual seems to have caught something somewhat larger than a Sand-eel, and indeed puffins throughout their range have been known to bring more than 50 different species of fish to their burrows. Other important species include Sprat (*Sprattus sprattus*), Atlantic Herring (*Clupea harengus*), Whiting (*Merlangius merlangus*), Capelin (*Mallotus villosus*) and Atlantic Cod (*Gadus morhua*). Puffin chicks are remarkably good at eating fish that at fish might look too big for them. **[BELOW]**

Much is made of the number of small fish that a puffin can hold in its bill at the same time. The record happens to be 82, all of which were actually tiny larvae. To a chick, however, it isn't the number that counts, but how big they are and how often the meals are brought in. Nonetheless, it should be recorded that most servings are modest, such as this batch of 8–10 Sand-eels. **[RIGHT]**

The iconic image of a puffin is of one on land with a beak full of fish ready to deliver to the young. This is the photograph that everybody wants to take, and the one that everybody recognises. It is in every sense, though, only a snapshot of time. Puffins carrying fish are generally in a hurry to get down their burrow, so that gulls cannot steal their catch. They only visit their youngster a handful of times a day. And the stage of fish delivery only lasts for about 40 days of the year, after which time the young leaves the nesting burrow on its own. For this reason, despite the familiarity of the scene, few people have witnessed it.

Nevertheless, the importance of the deliveries is such that the puffin has special adaptations for the seasonal chore of carrying fish. Both the bill and the tongue are fitted with small backward-pointing projections that hold the catch in place when the puffin is flying back to its burrow. This is the only time in the year that the puffin will need them; the rest of the time it will simply swallow any fish that it finds.

A puffin chick does not usually get any food on the day of its hatching, but instead feeds off what remains of its egg yolk. After this, however, it relies on deliveries. For the first few days, when the adult arrives with a meal it stands next to the chick in the nest chamber and calls softly to it, whereupon the chick will take the first few fish bill to bill. The adult drops the rest and, as the chick gets older, it will begin to leave fish close to the burrow entrance. For most of its early life the chick will remain in the burrow and has an instinctive aversion to daylight, something that only wanes when the youngster is close to fledging.　　　　　　　　　　　　　　›

For those who study, watch or photograph puffins, this is a moment to savour. The first individual of the season bringing in a bill-full of food is a sure sign that its egg has hatched. On the whole, bird colonies are places where breeding is fairly synchronous, so it won't be long before all its neighbours and colleagues will be doing the same. [LEFT]

On the whole, the adults bring in slightly smaller fish when their chick has just hatched, but the size soon levels out. [OVERLEAF]

When it is young, one parent remains in the nest to brood the chick, while the other goes out to fish. The parents take the duty in turns, but overall during the nestling period the female brings in slightly more than the male. Fish are brought in much less often than you might expect, usually between 3 and 11 times a day, with the record being 28. The average tends to be between 4 and 6 a day, although this varies between years and colonies. On a given day, the parents start their provisioning with a bang in the early morning, with a long pause in the middle of the day and a small increase in the evening. No food is brought in at night.

On average, the weight of each load is 8–12g of fish, with once again slightly more delivered in the early morning. The number of fish famously varies, but it is the weight and nutritional value that counts. Over the years, yields have gradually got smaller in most puffin colonies in Europe.

The most important fish delivered to most puffin colonies is the Sand-eel (*Ammodytes*), which usually outnumbers everything else. However, over 50 species of prey species have been recorded, and there is evidence that some, such as sprats (*Sprattus sprattus*), are caught selectively because of their high nutritional value: more sprats are caught than would be expected in relation to their abundance. There is an impressive Who's Who of exotic sounding names on the overall menu (fancy Jelly cat *Lycichthys denticulatus* anyone?) And just occasionally, puffins bring in crustaceans and even squid instead, which they are probably used to finding for themselves in the non-breeding season. ▪

One of the reasons why puffins can bring in a lot of fish at once is that they tend to attack schools of fish, rather than just singletons. Contrary to what you might expect, each fish is captured individually by sight, even in the gloom of deep water. Puffins tend to go for the larger fish within a school, estimating this by girth rather than length. **[RIGHT]**

Puffins are unusual in being to deliver so many fish at once. Their relatives the Guillemot and Black Guillemot, for example, only bring in one at a time, as do terns, for example. The trick is that both the tongue and the bill hold the fish in place. Once a fish is caught, it is lodged at the base of the bill crosswise, usually with the edge of the bill lodged into the gills. It is held by the tongue, which allows the bird to go ahead and catch another fish in its jaws. Each time a fish is added, it is manoeuvred into place, while the other fish adhere to the sides of the bill because of small projections on the edge of the mandibles. These projections are backward-pointing. The outer third of the puffin's tongue is stiffened to hold food, while the inner two-thirds is relatively fleshy. **[ABOVE]**

By far the most important prey item brought in by puffins for their chicks is the lesser Sand-eel (*Ammodytes marinus*); this individual has caught six. This slim-bodied fish spends much of its life in sand on the seabed, but during the summer it gathers in large shoals to feed on plankton, at which time it is most vulnerable to puffin attack. [LEFT]

Some authors have stated, a little mischievously, that puffins always arrange their fish alternately in the bill, one facing left, the next one right, left again and so on. This certainly happens sometimes, as this bird shows, but in fact it seems that the orientation simply depends on how the fish was caught, and the birds don't arrange the fish before setting off back to the colony. Incidentally, some fish are delivered when still alive, suggesting that the parents have not flown far to get them. [BELOW]

This puffin, beak full of food for his young, announces his return to the colony at Sumburgh Head by raising his wings. [LEFT]

LEAVING THE BURROW

Young puffins, or 'pufflings' remain in their burrows for about 38 to 44 days until they fledge. However, this figure varies enormously according to how much food is delivered. In periods of shortage it can go up to about 50 days, and in very extreme circumstances even longer; one unfortunate chick in Newfoundland finally left the nest 83 days after hatching. Or perhaps it was fortunate, because severely delayed chicks usually die.

Life in the burrow is more active than you might expect. The chick doesn't just sit passively waiting for its handful of feeds a day, but instead walks around and will play with any debris it can find, such as nesting material. It will also attack any intruder without hesitation, even if it is an adult that might have mistakenly wandered in. The reason for this is interesting: studies have shown that you can easily swap puffin chicks between burrows, and the adults will still come in to feed them. That suggests that the parents don't recognise their chick, neither by sight nor by sound, but will blindly feed a strange youngster. Thus any chick in a burrow needs to hold its ground in order to survive. In theory a neighbour could take over its nest chamber and steal its food. ›

This fledgling is making a short visit outside the burrow in the company of a parent. It is exercising its wings, which are fully grown and feathered, and it could well be that, in a few hours time under cover of darkness, it will leave the colony and swim out to sea. Despite the display of affection here, adult puffins don't recognise their young chicks, which can be swapped between burrows without any distress of change in provisioning rate. Who knows, though, whether adult and young learn to recognise each other during encounters like this? **[RIGHT]**

Pufflings are slightly smaller than their parents, have dusky cheeks and, as you can see here, also lack any colour on the bill. When seen on the sea, they are scarcely identifiable as puffins. Notice that, despite its tender age, this fledgling is already familiar with the billing display. However, the ceremony is brief; after a quick flap of the wings, this juvenile dived back into its burrow. **[ABOVE]**

The survival of young fledged puffins appears to be remarkably high, with various measurements between 85 per cent and 93 per cent survival per annum. **[RIGHT]**

The fledging of pufflings is serene by comparison. For a start the youngsters don't leave the burrows – where they are pretty safe, remember – until they have reached about 70 per cent of the adults' weight, and are already equipped with strong, well-developed wings. They also leave at night and, in contrast to the case with most bird species, do so entirely on their own, with no adult present. In many ways their departure could hardly be more discreet, and although many pufflings within a given colony leave at the same time, this exodus does not provide a bonanza for predators.

The departure is quick and emphatic, and by first light the youngster is typically well out of sight of land. The fledgling has no further contact with its parents at all, and quite possibly never sees them again. They feed the chick right up to the night of its fledging, and not uncommonly appear the following day with food unaware that it has already gone. The fish are left at the burrow entrance and remain there, a cue for the parents to cease their provisioning. For their part, the adults often stay around the colony for a week or two after their chick has left. The fate of their progeny is out of their hands. ▨

At the end of summer, small numbers of first year birds turn up at puffin colonies for an 'anniversary visit'. Having spent a year at sea, and having shunned the land in the early stages of breeding, the birds appear on the edge of the colony and loaf about. Many youngsters eventually settle to breed at their natal colony, but by no means all do so – otherwise no new colonies would ever be formed. [OPPOSITE]

A puffling above ground is an unusual sight. When close to fledging, the restless youngsters often venture into daylight for short periods of time. This probably helps them get an idea of what their colony looks like, enabling them to recognise it in future summer seasons. The chicks leave the burrow for good only at night, and will not return until just over a year later. [LEFT]

PREDATORS AND STRIFE

Compared with many birds, a puffin lives for a long time. Once it has become an adult, the average individual can expect to live for around 18 years, far longer than a land bird of similar size. A wood pigeon, for example, which is of similar bulk to a puffin, has a life expectancy of only three years. Furthermore, the annual survival rate for a puffin is 90 per cent, meaning that nine out of ten puffins present at the breeding colony in one year will return safely the next. The figure for land birds rarely reaches 50 per cent. Furthermore, the oldest puffin yet recorded had reached the impressive age of 41 years. That isn't to say that puffins don't succumb to all kinds of mishaps, diseases and predators. They do, and this chapter deals with some of the things that can go wrong.

The most dangerous place for a puffin to be is at the breeding colony. Puffins are seabirds for which the land is not their medium of choice, nor the one to which they are primarily adapted, making them far more vulnerable there than they are on the sea. Furthermore, to breed successfully puffins must make repeated visits to a single burrow, and this regular commuting opens up opportunities both to predators and to food-pirates, which quickly take advantage of the comings and goings of hard-pressed parents. Gulls and skuas can be the bane of a puffin's existence during the breeding season, as well as being a genuine menace.

Puffins select islands in order to be free of ground predators, but of course sometimes mishaps occur and the balance changes. This seems to happen most often in the case of rats (*Rattus norvegicus*), which if they are introduced to an island, eventually wear down populations by preying on eggs or young. Puffin colonies can actually be wiped out by rats, whereas the sheer ledges used by guillemots and kittiwakes (*Rissa tridactyla*) afford better protection.

A puffin eyes up the hulking body of a nearby Herring Gull (*Larus argentatus*). Gulls sometimes stand in the way of puffins entering their burrows, provoking a stand-off that may last for some time. Herring gulls only rarely attack and eat puffins, but they are frequent food-pirates. **[RIGHT]**

An interesting case is that of Ailsa Craig, an island in the Firth of Clyde, in west Scotland, which once had a substantial puffin population. Rats fleeing from a shipwreck in 1889 came ashore on Ailsa Craig and within a few decades the puffin population collapsed, virtually to nothing. An eradication programme in 1991–92 removed the rats and, after a number of years puffins returned in small numbers. There are now about 130 pairs there, and the population is set to increase. Similar destruction by rats has been noted on a number of islands, most notably Lundy, an island off north Devon that is actually named after the Norse word for puffin, which goes to show that it must have had very large colonies once. However, in 2000 there were hardly any left, until an eradication programme targeting both brown and black rats (*Rattus rattus*) stabilised the puffin population and allowed it to increase.

Mishaps do occur on the sea too, of course, but they often go undetected. Nevertheless, from time to time large numbers of auks, which include puffins, can be found washed up on beaches, a phenomenon known as a 'wreck'. Birds involved in wrecks are often emaciated and the bodies tend to be washed up after a period of particularly harsh, prolonged stormy weather. From this we can conclude that puffins at sea can die of starvation from time to time, perhaps when the water is particularly rough and their usual surface-diving strategy is somehow impaired.   ⟩

They might look harmless, but these Great Black-backed Gull chicks could grow up on a diet of puffin flesh. **[LEFT]**

The puffin's number one predator, apart from man, is the Great Black-backed Gull (*Larus marinus*). The gull invariably chases the puffin in the air and grabs it from behind. Some individual gulls are puffin specialists and spend the greater part of the breeding season hunting them.   **[ABOVE]**

Heavily outnumbered, a pestered puffin holds on grimly to the single fish still in its keeping, despite the attentions of three Black-headed Gulls (*Chroicocephalus ridibundus*). This is an all-too frequent experience for parent puffins on Inner Farne, Northumberland. Here the burrows are close to a Black-headed Gull colony, and their neighbours take advantage of frequent free meals when adults return with food for their young. No individual can hope to hold on to its catch when the mob descends, so most arrivals keep a close watch before making a final dash for its burrow. **[LEFT]**

Great Skuas frequently eat puffins, although it is clear that some individuals are more prone to do this than others, and indeed become specialists. On Hermaness, where this picture was taken, only about 5 per cent of the skuas regularly attack seabirds, but their toll is still substantial. In one year at Hermaness there were 7000 to 8000 auk deaths attributed to Great Skua predation. The puffins often suffer most when fish stocks are low, forcing the skuas to switch away from their preferred diet. [ABOVE]

In contrast to Great Skua, Arctic Skuas (*Stercorarius parasiticus*) offer no direct physical threat to puffins, but they are far superior food-pirates. When a puffin approaches its breeding colony, these hawk-like birds usually ambush it well out over the sea, beginning to chase it from behind. They are highly agile and manoeuvre in intimidating fashion very close behind the auk, rather like a tail-gating lorry on a motorway. This is alarming for the puffin, and for an Arctic Skua the ideal result outcome is that the puffin simply drops the fish load it is carrying, whereupon the skua will collect it from the surface of the sea. Interestingly, when in flight puffins are much more reluctant to hold on to their loads than they are when being harassed by gulls close to the burrow, and chases may go on for some time. [OPPOSITE]

One would not instinctively expect a bird that eats fish in the sea normally to have trouble finding food, but it is clear that shortages do occur from time to time, and these impact most obviously on breeding success. In some years puffins breed in reduced numbers, and in others they barely breed at all, or they do and large numbers of their chicks starve. It is usually impossible to work out why shortages occur, since the sea is a complex and poorly understood set of ecosystems.

Of course, one factor that has had an effect on puffin populations for centuries has been human activity. Up until recently this hasn't been too much of a drain on the species. In times past, almost throughout its range the puffin was hunted and eaten by humans, and small numbers are harvested even today, as are puffin eggs. But this has always been difficult and hard work, and the flesh has never been rated especially highly, so it is unlikely that direct human persecution ever made much of a difference.

A puffin glances upward as a gull wheels over. Puffins are among the smallest seabirds in Britain, and vulnerable to attacks from the sky. Their choice of islands and steep slopes for breeding usually keeps them safe from ground predators such as Foxes (*Vulpes vulpes*). **[ABOVE]**

It's only a petty irritation when a lamb gets in the way of a nest-building puffin. In some areas, however, livestock occasionally trample over burrows and cause them to cave in. **[RIGHT]**

Today, though, that has changed. Direct persecution has given way to indirect pressure on puffins in the form of pollution (and probably overfishing and climate change, too). And being indirect, the dangers are insidious, poorly understood and difficult to measure. Some puffins die from oil pollution, sometimes when the feathers get coated, and sometimes from poisoning. All manner of chemicals are released into the sea, with unknown effects. Various kinds of plastic waste, such as bags, could also harm puffins. But puffins are small birds with remote wintering ranges that are hard to study at the best of times. Their fate in the seas around us is dangerously hard to monitor and predict.

Having said that, nobody is yet suggesting that the puffin is a threatened species. Its population is declining, but not yet steeply. The remoteness of its habitat, and its personal longevity, is for now a buffer against disaster. For now, it is still possible to visit a colony of this species, and witness its magic. ▦

Puffins aren't immune to northern Scotland's infamous midges. Summer evenings on Hermaness can be almost intolerable. For a puffin, the fleshy parts around the eyes and the bill are the most vulnerable, which is why this puffin is scratching so hard. Spare a thought, too, for the photographer being eaten alive at the same time as his subject. **[LEFT]**

A more serious and long-lasting irritation than midges or mosquitoes can be caused by ticks and lice. This individual is particularly badly affected by the former. Birds with such heavy infestations often appear weak and listless, and may well suffer from infections carried by the parasites, or even from anaemia. Fortunately, ticks eventually drop off. **[BELOW]**

INDEX